BEYOND EINSTEIN

By

Lenard Metzger

Other book by Lenard Metzger

COMMON SENSE COSMOLOGY

ID: 332616
lulu.com

BEYOND EINSTEIN

Library of Congress Control Number: 2008903425

Published by
Lulu Enterprises
Morrisville, NC 27560
United States of America

ID: 1681945
lulu.com

ISBN 978-1-4357-1461-8

PREFACE

How could people have believed that the world was flat and that if you sailed a ship to the edge your ship would fall off? Why didn't they realize that all the water in the seas would pour off the edge and the ship would have nothing to sail on?

After it was generally accepted that the world was round, there was this idea that the Sun and the stars all revolved around the Earth. And even when it was finally understood that the Earth revolved around the Sun, the belief was that the stars were on the surface of a crystalline sphere that rotated around us. How could people have believed in crystalline spheres with nothing beyond them?

I have tried to visualize this, but every time I reached a mental barrier in space I continued outward from the other side of the barrier. I have concluded that there is no end to space.

A theory of black holes shows mathematically that all the mass of a large star can shrink down to zero size and still have an extremely large mass. Similarly, a theory of the beginning of the universe assumes that all of the mass of the eventual universe started out with zero size. How can people believe this?

The above is a condensed version of some of the preface of my previous book, "Common Sense Cosmology". It was a personalized presentation. It consisted of factual chapters, but also included fanciful sections. I had hoped that it would be understandable to non-technical readers. This was not accomplished very well.

In this book, I have eliminated most of the superfluous sections. I am concentrating here on my factual, if somewhat speculative concepts. I have added chapters with the ideas that were worked out during the interim period. I intend to use a more rigorously technical presentation.

It has become obvious that my ideas are in conflict with some of the theories of Albert Einstein. I will describe where I think Einstein erred. Not being restricted by his theories allowed me to develop new solutions to several basic problems. I understand that this is presumptuous, but I have no other choice but to have my say.

By the definitions and equations that Einstein used in his theories of relativity he prevented himself from solving the problem he spent most of his life working on. That was to unify the theories of electromagnetism and gravity.

I will present, in this book, a theory that gravity is the result of a repulsive force imparted by particles having a velocity greater than that of light. These particles also determine the speed of light. I call these particles gravitons, and later describe what I believe them to be. This is not the conventional definition of gravitons.

BEYOND EINSTEIN

CONTENTS   PAGE

Introduction	7
Space-Time	9
Starlight	11
Universe	15
Multi-Verse	21
Gravity	27
Weight	33
Light	37
Radio	43
Hydrogen	51
Stars	63
Black Holes	85
Big Bang	93
Gravitons	95
Motion	101
Faster-Than-Light	103
Conclusions	109

INTRODUCTION

I consider cosmology to be the study of everything, including our universe and its place in the cosmos. We will consider our universe, as it appears to exist at this time, and then look at where it came from and where it is going. To be able to do this we will have to establish some fundamental assumptions.

I believe that it is safe to assume that our universe began with something that was somewhat like what is called the "Big Bang". An expanding universe was postulated based on the red shift of various stars and galaxies, as a function of their distance from the Earth. Hubble and others quantified this relationship. The age of the universe was estimated at between 10 and 20 billion years. This was arrived at by projecting the estimated rate of expansion (the Hubble Coefficient) backward in time to where all the observed bodies seemed to have coincided.

The present theory of the earliest time of the universe is that the temperature was so high initially, that only gravitational energy existed. As the expansion continued and the temperature dropped, successive energy régimes occurred; strong force, weak force and electro-magnetic force. Then subatomic particles appeared, followed by the first atoms, hydrogen, deuterium and helium. Over vast periods of time, volumes of the hydrogen gas converged, becoming stars that combined into galaxies.

The above concepts will be the starting point for my new ideas. The approach that will be taken is to use the least amount of mathematics possible and use simple logic to describe my vision of how the cosmos works.

I will assume most generally accepted astronomical and physical facts as givens. Conclusions that have been reached as the result of manipulations of derived equations will be questioned and subject to re-interpretation. There are a number of prevalent theories that are not so easily understood or visualized. In some cases they are not even logical. In the following sections I will address these cases in detail and offer alternative explanations.

What I question is using a mathematical model of a phenomenon that requires adjusting the coefficients of the equations of the model, to give the desired answers. Even worse is extending the predictions of the model well beyond its normal range of data. Such a model could have no connection to reality. What I have tried to do is show the possible, underlying cause of a phenomenon, using the simplest, physical explanations. I believe there is an ultimate reality that should be sought.

SPACE-TIME

All that the cosmos consists of is matter, energy, space and time. I believe that the first three of these things are fundamental and that the last (time) is not.

All of the particles and bodies in space are matter. Energy when applied to matter causes it to accelerate and achieve an increased velocity. Space is the distances between bodies and can be assumed to extend beyond the most distant body.

Time is determined by making use of the above statements. Time is the duration of the motion that a body undergoes at a given velocity to travel a distance in space. It is given by the ratio of distance to velocity. Distance and velocity are observable, independent variables. Time is the dependent variable.

We can see two objects at a distance apart. We can move from one object to the other, slowly or quickly. We cannot see time. We can use the duration of the day to give us a time reference. We can walk, putting one foot against the other for 5280 steps and define this distance as a mile. We can walk slowly, back and forth along our mile, and count the number of times we do so, for a whole day. The number would be our velocity in miles per day. If we walked faster the number would be higher and we would calculate a higher velocity.

We can mechanically or electronically divided the day into smaller parts, hours, minutes, seconds, and microseconds. We can use these to express the ratio of distance to velocity, for very short distances or very high velocities. Sixty miles an hour is still the same velocity as a mile a minute.

This is a key concept when we consider the travel of light over a distance. The velocity of light has been defined by Einstein to be a universal constant, the maximum velocity physically achievable. The vast distances in space are described in units called "light years". (These years are Earth years, not Mars years).

The source of ones perception of time is the brain. Within the brain, the distance between two adjacent nerve endings and the velocity of the neural transmitter molecule that travel across this distance determine the quickness of ones perceptions and reactions. The basic clock rate of our thought process depends on distance and velocity.

Einstein developed his theories of relativity on the basis of velocity and time being independent variables, which made distance the dependant variable. His four dimensions are the three spatial dimensions (X, Y and Z) and a distance dimension determined by the product of velocity and time. This ultimately led to one of his conclusions that had time varying as a function of velocity relative to the speed of light, independent of distance. It also led to the concept that space itself could expand or contract. I believe that having time as an independent variable was one of the factors that led to these illogical results.

My belief is that the observed motion of light and matter can be explained by their interactions with gravitons. It is gravitons that produce the effects of gravity. There is no need to be bound by Einstein's relativity theories. This allows space to be considered as invariant, infinite in extent. Also, this allows time to be considered as merely the infinite duration of the total of all events.

STARLIGHT

The Sun and all the stars we see with our naked eyes are part of what is called the Milky Way Galaxy. Our galaxy is composed of billions of stars arranged in a flattened disc shape. Our Sun is located about half way between the center and the edge of the galaxy. When we on Earth look toward the center of the galaxy the stars appear to get closer and closer together, until they seem to become a solid band of white, the Milky Way. If we were to look between the stars through a large telescope into the far reaches of the universe we would see many other galaxies like the Milky Way. There may be billions of them.

The measurement of the wavelength of light is called spectroscopy. It has been very important in the understanding of stars, galaxies and the cosmos. Many wavelengths are mixed together in the light coming from the Sun. The early scientists that studied this sorted them out with some difficulty. Laboratory tests under controlled conditions helped. After establishing the nature of these spectra they were able to detect the shift that had occurred in the spectra from the light of distant stars and galaxies.

Measuring the spectra of the light emitted has led to the knowledge of star size, temperature, and direction and speed of motion. The hotter the star the higher the energy of the excited hydrogen in its atmosphere and therefore the bluer its radiated light due to the higher energy photons.

It should be noted that there are two kinds of spectra. One consists of the bright lines emitted by the excited atoms. The other spectrum consists of dark lines in the almost continuous bright light from the star. This spectrum is produced when cooler atoms, selectively absorb photons passing through the outer region of the star.

Color is an important characteristic of light. It is what we perceive in the visible portion of the spectrum of light. But we should consider it as the wavelength of light, whether it is infrared, at a longer wavelength than visible light, or ultra-violet light, at a shorter wavelength than visible light. As an electro-magnetic wave, a specific "color" of light oscillates at a specific frequency as it travels along at light speed. The distance that it travels in one cycle of oscillation is the wavelength.

As an example, the wavelength of red light is almost a micron, which is a millionth of a meter. Since the speed of light is about 300 million meters per second and the frequency is the speed divided by the wavelength, this gives a frequency for red light of about 300 trillion cycles per second. This is why light is usually characterized by its wavelength.

An important phenomenon to understand is what is called the "red shift". This is used to measure the velocity of a star or galaxy relative to the Earth. The greater the velocity of separation, the more the characteristic blue light appears to be shifted towards the red end of the spectrum.

To understand this, consider a star moving away from us at a tenth the speed of light. Think of the first cycle of the light at the beginning of one second, moving the known distance of 300 million meters during that time. The last cycle, emitted after the star had traveled one second, would be 330 million meters behind the first cycle. Dividing this slightly greater distance by the known frequency of the light results in a slightly longer wavelength.

Therefore, when we detect this radiation, it appears shifted towards the red end of the spectrum.

Another aspect of this is that the same effect would occur if the distant object were stationary and we were moving away from it at some fraction of the speed of light. During a one second period, that we would measure the light, slightly less of the radiated wave would pass us. This would result in a lower number of cycles per second being detected. This results in a redder, longer wavelength light observed.

UNIVERSE

Observations have been made of the light from quasars, estimated as having been emitted about 12 billion years ago. These emissions are thought to have taken place only one billion years after the Big Bang. Repeated observations of these objects indicated that the rate of change of their red shift was increasing. It was concluded that the expansion of the universe was accelerating. A negative force of gravity was postulated to account for this occurrence.

However, studies of the distribution of galaxies, as a function of distance from us, indicate that space is essentially flat. I will take a Newtonian view of the universe and assume space is linear and not expanding. I believe there is a different way of explaining an increasing red shift in the above example.

Instead of the usual assumption that all galaxies are moving away from us with velocities that increase with distance, assume that astronomical bodies move away from where the Big Bang occurred with a wide range of radial velocities. Also, assume that the Milky Way galaxy has about an average radial velocity.

The quasars may have been moving very slowly, either toward us or away when they emitted their light. If they had a decelerating outward motion or an accelerating inward motion, the result would be an increasing rate of red shift change.

The amount of red shift of light we observe from distant objects depends on the rate of increase of the distance between them and our galaxy. There are several ways that we may observe the same degree of red shift.

The distant source of light could be moving in a direction opposite to the motion of our galaxy. The observed red shift would be a function of the sum of their velocities.

There could be an object moving in the same direction as our galaxy, but much slower. It would then be closer to the center of the universe than our galaxy. Another object could have started in the same direction as our galaxy but with a greater radial velocity. It would then be farther away from the center than our galaxy. In these cases the red shift would be directly proportional to the difference in velocities between them and us. The following figure illustrates these things.

RADIAL DISTANCES IN BILLIONS OF LIGHT YEARS

⟶ VELOCITY VECTOR

------> RELATIVE VELOCITY

GALAXY POPULATION DENSITY (CROSS SECTION)

Figure 1

Assume that our galaxy is located at point A. Its radial velocity is shown resolved into vertical and horizontal components. Point A is shown at a distance from the center (The Big Bang) that is near the peak of the galaxy population density

A galaxy at point B had started with a larger radial velocity than our galaxy. The vertical velocity component of the B galaxy was greater than that of our galaxy at the time that B emitted the light that we now receive. The red shift that we observe is a result of the difference between the two vertical velocities.

The galaxy at point D started out with a smaller radial velocity that our galaxy. Its vertical velocity was less than that of our galaxy. We were moving away from D at the time D emitted the light we are just detecting, with the observed red shift as a result.

Another possibility is to consider a galaxy that started with about the same radial velocities as our galaxy but at different azimuth angle. This is shown in the above figure as the galaxy at point C. This galaxy and ours are both moving away from each other. The red shift results from the sum of the horizontal velocities. This red shift would be considerably greater than for the other two cases, even though all are at the same distance from us.

This could account for the almost two to one variations in the measurements of the Hubble Coefficient by different observers and the various estimates for the age of the universe of from 10 billion to 20 billion years.

It is reasonable to assume that other universes preceded ours. While the course of the expansions and contraction of the prior universes cannot be known precisely, we can assume that they ended with conditions that led to the initiation of the next universe. It is likely that there was sufficient matter that converged to a central concentration of mass that produced a "Big Bang".

If a model of an expanding and contracting universe were made it might show that by the time an appreciable amount of matter had accumulated at a central point, much of the remaining matter would be relatively close, consisting mostly of galaxy-weight black holes.

Assuming that the concentration of exotic matter in the central mass had reached a critical density or state, an impact from another massive black holes may have triggered the Big Bang. I believe that the Big Bang was a fact of nature that happened as a naturally occurring event. Also, if something happened once, then the conditions that led to it can happen again.

When the detonation occurred and the expanding plasma reached the point of atomic particle formation, they could have started responding to the mass of the surrounding black holes. Some of the high velocity plasma would have been trapped into orbits around the black holes, perhaps producing what we detect as quasars. The disruptions in the outward flow of the plasma would have resulted in variations of density and velocity. This could account for the eventual clusters of galaxies and the voids between galactic clusters.

The momentum of the expanding plasma would interact with the momentum of inward moving black holes. The results could be a slowing of the inward motion or a reversal of the motion of the black holes. The massive black holes would probably resume their inward motion with some exhibiting quasar activities.

A black hole with only the mass of a single star would be capable of attracting a concentration of expanding gasses but would be carried along with the gasses. This might explain the initiation of galaxies with black holes at their center.

MULTI-VERSE

The following is another possible explanation for the change in the rate of red shift for the most distant objects we observe. Massive objects lying beyond our universe perhaps could affect the outermost objects in our universe. Perhaps other universes surround our universe.

The prefix "uni" in universe comes from the Latin for "one" but in the word "universe" it is considered to mean "the one and only". I believe it is conceivable that our universe may be only one of many simultaneous universes. I would like to call this the "multi-verse" consisting of all the universes, since "multi" is the prefix for "many". This is also my understanding of what the "cosmos" can be considered.

My ideas about simultaneous universes will need some explaining. I believe that space is endless, infinite in extent, and linearly three-dimensional. I also believe that time is also infinite in extent, without a beginning or end.

Assuming that the age of our universe is 20 billion years, at most, then the light emitted at the time of the Big Bang has now reached the surface of a sphere with a radius of 20 billion light years. The sphere containing any appreciable matter, such as high velocity gasses, probably has a slightly smaller radius.

We can now picture our expanding universe as an isolated sphere floating in infinite space. What is wrong with this picture?

Why should there be only one unique universe? An infinite space has room for an infinite number of universes.

I believe that the existence of any matter and energy at all, means that there can be an infinite amount of matter and energy in existence.

I have pondered on the source of all of this and have come to a conclusion. If one believes that everything was divinely created, then that is the answer. The only alternative that I can see is that all matter and energy has always existed. I often come back to thinking; "but where did it all come from originally?" I have to remind myself that it always existed.

If there are other universes surrounding ours, how could they be spaced around ours? With the simplifying assumption that they are all spherical and about the same size, the densest configuration would be for about twelve other universes to surround ours, essentially in contact. This would place ours in a somewhat favored position, which I don't think is warranted.

Or other universes could be located randomly around ours at varying distances and orientations. I think that over a large number of iterations, of interactions and big bangs, the random configuration would come to have a fairly regular arrangement.

A simple hypothetical assumption would be to place our universe in a cubic lattice. This configuration could be repeated indefinitely in all directions with none of the universes in a favored position.

We could picture our universe at the central vertex of the eight surrounding cubic volumes. This closest cubic array has a total of twenty-six other universes. This is shown in the following figure.

Let us consider the dimension of the side of the cube to be some unit to be determined.

Figure 2

There would be six universes at this closest distance away from ours. These six universes are those along the X, Y and Z-axes. There would be twelve more universes centered at a slightly larger distance away, the square root of two times the unit distance. Finally, there are eight more centered at the square root of three times the unit away.

The gravitational effect of these surrounding universes should be calculable. At the center of our universe they will produce equal and opposing fields from each opposing pair of universes. This effect could represent what has been proposed as the Higgs field.

As an outermost galaxy in our expanding universe gets closer to one of the other universes and farther away from the opposite universe, it may be effected by a difference in the gravitational effects, and undergo acceleration toward the closer, adjacent universe.

From the acceleration that has been observed, an estimate of that unit of distance between universes might be attained. (The above assumed that gravity is an attracting force, which I no longer believe.)

Although the surrounding universes may be at different stages of expansion they eventually could reach a point were two or more of them start to overlap. This might produce a situation where in that region the concentration of mass would reach a larger density and start to compress toward a new "Big Bang". Obviously this event would occur in the regions between groups of universes. In this way, universes that are expanding towards oblivion can give birth to new universes by interacting with one another.

That all the universes would begin at about the same time is based on the following assumption. Any expanding universe would spread an eighth of its total mass into each of the adjacent lattice cube volumes. Each of the universes at the eight corners of each lattice cube would do the same. The largest concentrations of mass would eventually occur in the center of each cube. This is where the total mass would again reach the critical value. The following figure shows eight universes expanding toward each other.

Figure 3

Groups of two or three of these universes can start accumulating large combined black holes where they overlap. Eventually all of these can collide in the center and trigger the Big Bang.

Figure

GRAVITY

Newton assumed that gravity is a force of attraction acting between material bodies. His approach to gravity is appealing for its simplicity but does not address what causes gravity. It only shows how the effects of gravity can be calculated.

Einstein, in his development of the theory of general relativity, also assumed that material bodies exerted a force of attraction that affected the space-time continuum. Neither man explained why this attractive force occurred. Einstein's mathematical approach leads to the concept of curved space-time causing the effects of gravity. By manipulating equations involving multiple variables and coefficients he was able to predict known experimental results. His explanation for gravity was that objects followed paths of least resistance in curved space-time. As a result of his equations, some explain the expanding universe merely as space expanding.

I have been struck by the need to have gravity act over great distances essentially instantaneously. I think that the gravity of the total mass of our universe may be felt far beyond the present radius of the Big Bang radiation.

What if at the time of the Big Bang, when the initial state consisted only of gravitational force, a gravity wave was radiated at a velocity far in excess of the present speed of light? What if gravity is always transmitted faster than light? Would it be an attracting force or a repelling force?

In my concept of multiple universes, I assumed that all the universes surrounding ours began at about the same time. All of these universes would have about the same total mass as each other. This is based on my concept of an accumulation of giant black holes reaching a critical mass and creating a Big Bang.

The gravity waves from all of these universes would be permeating our universe from all directions. The effect of this would tend to balance out on the large scale within our universe. However, if these gravity waves were repulsive, a large body could tend to shield a smaller nearby object from the waves coming from beyond the larger body. This would make the smaller body move towards the larger.

One theory of gravity assumes that gravity is transmitted by an intermediary particle called a graviton. It is said to be a mass-less particle traveling at the speed of light and having momentum. If gravity is transmitted by a mass-less particle, then there is no reason that at would be limited at the speed of light.

Now I am proposing that gravity is caused by gravitons transferring momentum to the atoms with which they collide. The greater the number of protons, neutrons and electrons in a body per unit volume, then the more the gravitons will interact with the body. I am assuming that these will be elastic collisions and not reduce the energy of the gravitons significantly. Many gravitons will pass through most material objects without any collisions. Those involved in collisions will be scattered over a range of angles.

Any isolated material body in space would be subjected to equal amounts of momentum from gravitons coming from all directions of space.

Therefore, there would be no net momentum added in any one direction. However, if a larger body were in the vicinity of a smaller body, it would shield the smaller body from receiving some of the gravitons coming from the direction beyond the larger body. This is shown in the figure below.

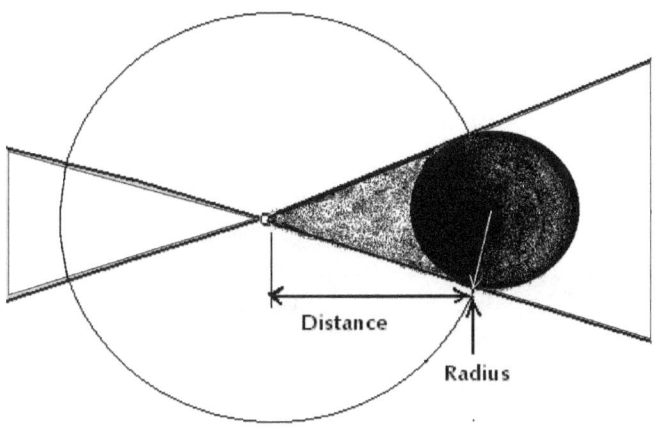

Figure 4

The amount of shielding would depend on the presented area of the larger body and its distance from the smaller one. It also depends on the density of the large body and how many of the gravitons it intercepts and how many it transmits.

Through the same solid angle that the large body presents, but in the opposite direction from the small body, the incoming gravitons are unimpeded. This results in a net difference in momentum applied to the small body. This will cause it to move towards the larger body with a specific acceleration.

The effect of the momentum will depend on the mass of the small body. For a given volume of matter, the greater the density, the greater the mass, and the greater the force applied to the body by the gravitons. Since force is equal to mass times acceleration, if the ratio of force to mass is constant, then the acceleration is constant. This is consistent with the gravitational coefficient of Newton.

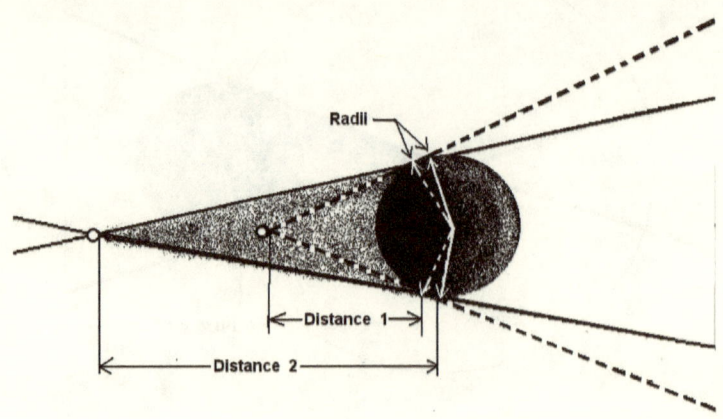

Figure 5

The solid angle presented by the larger body depends on the distance between the bodies. As shown in Figure 5, the solid angle will vary as the inverse square of the distance.

The inverse square relationship can be explained as follows: The presented area of the larger body is approximately pi times its radius squared. This will be a small fraction of the total area of a sphere centered on the smaller body, calculated for the specific distance to the large body. The area of this sphere equals four pi times the distance squared.

The fraction of the total radiation intercepted by the larger body is given by the ratio of these expressions. This is the radius of the larger body squared, divided by four times the distance squared. This simplifies to a constant divided by the distance squared.

This applies actually when the presented angle is small. As the smaller body approaches the larger body, this angle will become large and the relationship with distance will deviate from the inverse square.

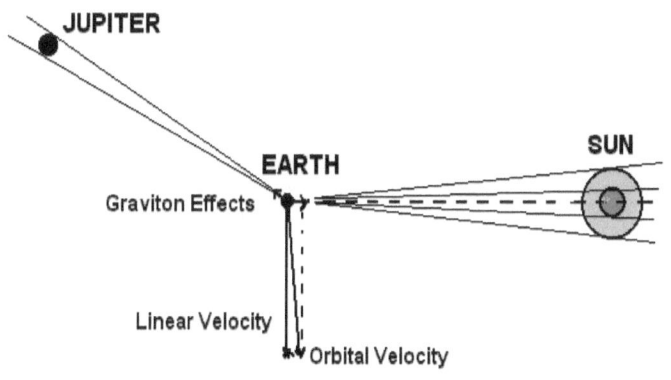

Figure 6

In the figure above an approach to calculating the effect of the Sun on the orbit of the Earth may be considered. The momentum applied to the Earth to change its linear velocity direction into its orbital velocity direction may be calculated. If the scattering coefficient of gravitons, for the various regions of the Sun and the Earth, can be determined, the net reduction of graviton momentum applied to the Earth could be calculated by integrating over their full volumes.

31

Similarly, the effect of Jupiter, at any particular distance and relative orientation might be calculated. The known deviation in the Earth orbit caused by Jupiter can be accounted for by the decrease in graviton flux caused by Jupiter.

Knowing the total mass of the Earth and the force necessary to accelerate it into a circular orbit should allow a calculation of the effect of the graviton flux coming from the opposite directions of either the Sun or Jupiter.

WEIGHT

Consider a small object that was caused to approach a large sphere until it stopped in contact with the sphere. This is shown in the following figure. The momentum applied to the small object will now cause it to press against the large object with a force that results in what we measure as its weight.

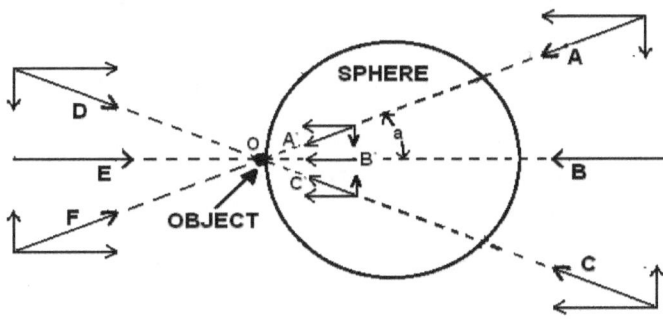

Figure 7

Assume that the sphere is the Earth. Any graviton passing through the small object from the right will have gone through the Earth for various distances. The horizontal path labeled B is the longest distance, through the diameter of the sphere. This path will cause the most reduction in the graviton flux. However, all of the exiting flux, labeled B`, will be along the horizontal direction and affect the measured weight.

For every path such as A there is an equal length path, such as C. Their path lengths will be shorter than the path B and will not reduce the graviton flux as much.

Their horizontal components will be reduced by a factor equal to the cosine of angle a. These horizontal components will add. The vertical components of A and C will be equal but oppositely directed and will cancel.

The horizontal components of the un-attenuated graviton fluxes, such as D, E and F, will add in the object O, and will be partially cancelled by the sum of the horizontal components of all the attenuated flux entering the object from the right.

The fundamental factors that will have to be determined for calculating weight are the following:

1. What is the un-attenuated, graviton flux density, coming from all direction of space?
2. What are the attenuation factors for various large bodies such as the Earth and the Sun? (i.e. Scattering coefficients, absorption coefficients, etc.)
3. How much force is transferred to the small body by the differential graviton flux?

In Figure 8 below, an approach to begin the calculation is shown.

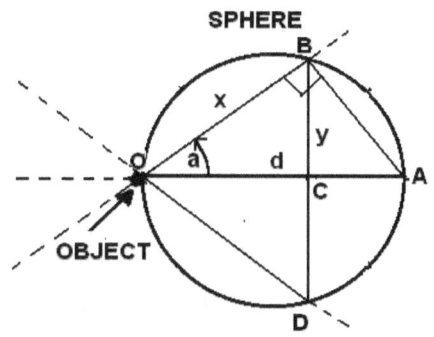

Figure 8

Simple trigonometry and calculus can be used to solve the flux components as a function of the angle "a". For a flux path through point B and going through point O, the distance "x", is given by the cosine of "a" times the diameter "d". This is because all triangles subscribed in a semicircle are right triangles.

The distance "y" going from point B to point C, is given by "x" times the sine of "a".
Therefore: y = d (sin a)(cos a).

All the paths that have the same value of angle a, enter the sphere through a circle whose circumference equals pi times 2y.

Each path through the sphere will have an attenuation factor that will vary as a function of the length of x. This function remains to be determined.

Put all of these together and add another factor of cos a, to convert to the horizontal components of the flux. Integrating the resulting expression from a = 0 to 90 degrees should give the total, attenuated horizontal component of flux entering the small object from the right.

Similar equations can be used to arrive at the total horizontal flux entering the small object from the left. Obviously, the attenuation factor can be omitted in this case.

An experimental start to determining some of the unknown parameters might be to measure the weight of a mono-crystalline object, with a simple geometry, such that its mass and total number of atoms can be calculated. Depending on the structure of the crystal, there may be differing probabilities of graviton interactions with its atoms, under various orientations.

From the probability of a graviton encountering an atom over the range of possible angles of incident, the average force that is transmitted to an atom might be obtained. Of course the ratio of the flux coming from the sky relative to that coming through the Earth must be determined.

On further thought, for a very small mass, the path length through it would be very short. Since this mass exhibits weight, the probability that gravitons interact with it is 100%. This means that the probability of a graviton passing through the Earth and not interacting with an atom is close to zero. This suggests that this attenuated graviton flux passing through the Earth may be neglected in the calculation.

There may be a minimum time duration required for sufficient gravitons to interact with a small body and arrive at a stable weight reading.

LIGHT

Trying to decide the nature of light has been a quandary for many centuries. In some ways photons of light behave like particles and in other ways they act like waves. For a hundred years it was assumed that light, as a wave, needed a medium in which to travel through space. This medium was called the "luminous ether."

A problem arose in trying to measure the velocity of light. It had been assumed that the velocity of earth traveling through the ether would produce a change in the speed of the light moving in the direction of the motion of the earth, as compared to the speed measured at right angles to this direction. Many attempts were made to detect this change in velocity, using instruments with ever increasing accuracy. No change in the velocity could be measured.

In a last effort to retain the concept of ether as the medium for electro-magnetic radiation, a theory was proposed that the units of length and time, used to measure the velocity of light, must change as a function of the velocity of light. The equations that described this effect are called Lorentz Transformations.

Einstein used these equations in his "Special Theory of Relativity". He applied them to all matter, except for light. He defined the velocity of light as a constant that is the maximum velocity that can occur. He assumed that this velocity was independent of the velocity of the source of the light. This was his attempt to reconcile theory with physical measurements.

Einstein expanded his use of the Lorenz Transformations. He stated that an accelerating object would have its mass approach infinity, as its velocity approaches that of light. Therefore, by this definition, a photon of light could have no mass.

This is in conflict with the known fact that light is able to impart momentum. Einstein knew this. His first paper, on the photoelectric effect, described how photons of light liberate electrons from the surface of a metal.

I assume that photons of light have mass. I also assume that the velocity of light is not the basic limit of the velocity of matter. Something else must be limiting everything, including light, to their observed velocities and masses. This led me to a new theory of gravity that I describe in this book. My concept of gravity is that it is caused by particles I call gravitons, traveling at "light speed." This has a direct bearing on my concept of photons of light being particles, having a finite mass, traveling at slightly less than "light speed".

One can visualize a photon traveling and being retarded by gravitons coming from the direction ahead. The photon would also be propelled by gravitons coming from its rear. If the photon reaches the velocity of the gravitons following it, the photon will no longer be propelled. Then the gravitons coming from the front will retard the photon until a balance is restored. This should occur at a velocity just below the velocity of the gravitons.

This process could produce an oscillation of the photon, back and forth along its direction of travel. Oblique interactions with gravitons could cause side-to-side motions of the photon.

The energy of a photon and the equivalent mass of that energy can be calculated. To derive the actual parameters of a photon, start with the equation for a quantum of light. The energy of the photon, E is given by the following equation: $E = hn$ (joules)

h is Planck's constant.

$h = 6.626 \times 10^{-34}$ (joule-seconds)

n is the frequency of the radiation in cycles per second, given by the equation: $n = c/L$

c is the speed of light ($c = 3 \times 10^8$ meters per second)

L is the wavelength of the light in meters.

Therefore: (Equation 1) $E = hc / L$ (joules)

Assume that the mass of the photon can be determined by using Einstein's equation:

(Equation 2) $E = mc^2$. (m is mass in kg)

Combining this with Equation 1 gives the following:
$mc^2 = hc/L$

Therefore: (Equation 3) $m = h/cL = (2.2 \times 10^{-42})/L$ (kg)

For green light, $L = 5 \times 10^{-7}$ (meters). Therefore, the mass of a green photon is about 4×10^{-36} (kg).

By a different calculation the mass of an electron can be shown to be about 10^{-30} (kg). Therefore, the mass of a green photon is about four millionth of the mass of an electron.

It seems reasonable that an excited atom could throw off such a small portion of an electron mass and charge. Traveling at the characteristic speed of light, it will provide the observed momentum of the photon.

For a possible explanation of the electro-magnetic characteristic of the photon, assume the following: In addition to its forward motion, the tiny charged particle vibrates in a plane transverse to the direction of propagation

The moving charge will induce a magnetic field that will tend to stop its transverse motion at the end of its excursion. The magnetic field will then start to collapse and cause the charged particle to reverse its sideways motion.

As the magnetic field reaches zero, the particle will have reached its maximum transverse velocity and start increasing the magnetic field. This will be of the opposite polarity, which should cause the particle to slow down and stop at the other end of its excursion. Thus, the cyclic transfer of energy between the charge and the magnetic field will continue. This is shown in Figure 9.

Figure 9

It is known that a photon with a higher frequency of oscillation has greater energy content. This suggests that a higher energy photon, with a larger mass and charge, will vibrate faster, with a smaller excursion.

It seems possible that the direction of the vibration could rotate around the direction of travel. This would account for circularly polarized light.

A characteristic of electrons that should be considered is that of electron spin. The spin of an electron around an axis through its center produces a circulating charge that causes a magnetic field that emerges at one end of the axis and returns at the other end.

If the small fraction of an electron that is expelled as a photon retains a portion of the spin, this could account for the magnetic portion of the electro-magnetic field of a photon. Its electric field should be oriented transverse to the magnetic field. This is shown in Figure 10.

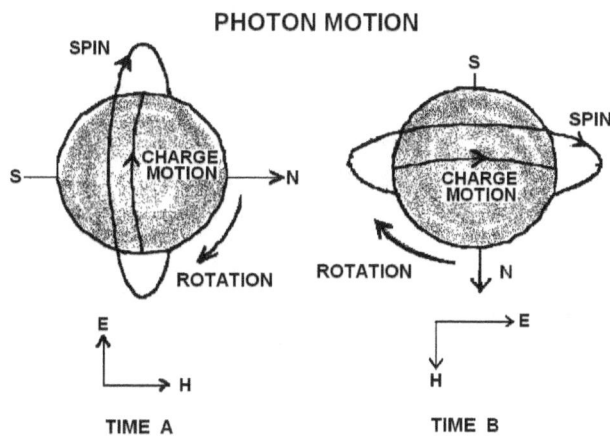

Figure 10

Figure 9 is a representation of the zero spin possibility. Figure 10 shows how the spinning charge might produce the electro- magnetic fields.

The direction of the charged photon path, in Figure 10, can be either across the page or into the page. A quarter turn of rotation is shown.

The fact that the polarity of the electro-magnetic wave reverses every half wave- length that the radiation travels could be accounted for by the particle axis rotating in a plane transverse to the direction of motion.

Another possibility is that the axis of spin could be oriented along the direction of motion. If the axis of spin were to rotate in the plane through the direction of motion, the wave front of the magnetic field would change polarity every half turn and go through a null at the time of reversal.

If the electric field were at a maximum at that instant, when the axis of spin is pointing along the direction of motion, the required alternating maximums of the electro-magnetic field might result.

I have described several possible modes of the photon particle motion to result in an electromagnetic field. I have learned that the value of spin assigned to an electron is ½. The photon has been assigned spins of 0 and ½. This might mean that both the modes described, that assumed spin and no spin photons, could apply.

RADIO

It has been generally accepted that all electro-magnetic waves are the same. They are suppose to be described by the Maxwell equations equally well for radio waves and light rays. How can my concept about photons be compatible with radio transmission and reception?

When an alternating voltage signal is applied to a dipole antenna, shown in Figure 11, the following events are said to occur. I will assume a typical broadcast frequency of one hundred megahertz. A half wave dipole would be about one and a half meters long.

A sine wave carrier is applied to the middle of the antenna as shown. At the maximum of the applied power the electron flow from one side of the dipole to the other will reach its peak value. A maximum magnetic field will be developed circling the current flow. This field will move away from the dipole at the speed of light.

As the applied power decreases and goes through zero, the electron current flow stops and no magnetic field is formed. At that instant the maximum number of electrons have accumulated at one end of the dipole and the corresponding number of positive charges have appeared at the other end. That is where the electrons have been depleted by the current flow.

Figure 11

As shown above, a shell of electric field lines suggests a connection from one side of the dipole to the other. This electric field also moves away from the antenna at the speed of light. It is obvious that in the next quarter of the cycle of voltage the electron flow will reach a maximum value going in the opposite direction. This will develop a magnet field circling in the opposite direction. This field will move off following the electric field, etcetera.

Electric power is put into the antenna and power is transmitted. The antenna is made of a good electrical conductor so that not much power is wasted in the electron flow, back and forth. It has been assumed that the excess input power is carried off in the electromagnetic fields. But what if there is the equivalent of photons that carry the power? What if the electric field lines consist of a line of tiny bits of electric charge emitted as the electrons move along the antenna? The bits would have a small motion along the field line. But their primary motion would be away from the antenna "at the speed of light."

There would be an elemental magnetic field ahead of them and another magnetic field of the opposite polarity developing behind them. After that magnetic field would come more tiny bits of charge with a slight motion in the opposite direction from that of the first particles.

In Figure 12 there is shown several of the cells, of the many, that form along the length of the antenna. These would interact in a complex interlocking circular motion while traveling outward.

The ovals represent the circulating magnetic flux. The electric charges move only slightly sideways, as they are moving outward at the speed of light. The pairs of charges are linked by their magnetic flux, as they tend to move tangentially past each other. The fluxes add between pairs and produce the alternating magnetic fields of the radiation (indicated by H).

Figure 12

Here I go inventing a new term. If there are radio photons, let us call them "rhotons." They consist of two charge bits per wavelength.

The power from this antenna is radiated in a pattern shown in Figure 13. This is a vertical cross section of the field strength of the indicated dipole. This pattern is toroidal but has also been called doughnut shaped. Looking down at it would show a circular pattern.

46

Figure 13

Three different receiving dipoles are shown at different distances and orientation relative to the transmitting dipole. Receivers at A and B are along the line of maximum field strength. B, being farther away than A, is at a point of lower strength. The receiver at C is off the maximum direction and has the same field strength as at point B, but at a closer distance to the transmitting antenna.

It is known that the field strength falls off with distance. With our assumption of elemental cells of oscillating charge and magnetic flux, how can the reduction in strength be explained? With the increasing area that the power is spread over with increased distance, the number of these cells in a unit area will decrease.

If the transmitting antenna is emitting radio type photons, what can be the mechanism? It is known that the electron flow on a transmitting antenna is mostly along the surface. This is called the skin effect.

Electrons moving in a metallic conductor, under the influence of an applied voltage potential, migrate by freeing valence electrons and by filling vacant valence sites. Perhaps this low energy transfer process would result in low energy quanta emitted from the surface atoms.

The common metals used in making antennas can be those such as steel or copper. They could be coated with silver. The atomic structure of these metals have from three to four completed electron shells surrounding the nuclei, with only one or two valence electrons in the last incomplete shell. The positive electric field of a nucleus is largely shielded from an incoming electron. So an electron interacts with the atom with a small velocity and a little energy exchange.

If the rhotons of the radio wave are comparable to photons, their characteristics should be calculable. Assume that Planck's equation,
$E = hn$, can be used to determine their energy. The 100 MH signal ($n=10^8$), would have an energy per rhoton of 6.6×10^{-26} joule.

Assume the transmitted signal power is 100 watts. That is 100 joules per second. One cycle of the broadcast signal takes 10^{-8} seconds. During one cycle of transmission, the amount of energy transmitted is 10^{-6} joules. The number of rhotons created in this period of time is 10^{-6} divided by 6.6×10^{-26}. This comes out to be 1.5×10^{19} rhotons.

To see how this output number relates to the power input to the antenna, consider that the radiation resistance of a dipole is about 70 ohms. The power input equals $I^2 \times R$, where I is the input current in amperes.

Solving for this current: gives the current as I=1.2 amperes. (An ampere is a coulomb per second.) A coulomb is a quantity of electric charge equivalent to 6.28×10^{18} electrons. Therefore, the input to the antenna is about 7×10^{18} electrons per second. During one cycle of input this is 7×10^{10} electrons. Dividing the above number of rhotons per cycle by this number of input electrons per cycle comes out to 2×10^8 rhotons per electron. This seemed like a lot to ask of each electron.

We will use the equation derived for the mass of a photon (m=h/cL) and apply it to a rhoton. In this case L=3 meters. Therefore, the mass of the rhoton would be 7×10^{-43} kg. Comparing this to the mass of an electron (10^{-30}kg), we get the ratio of 10^{12} rhotons per electron. The number of rhotons taken per electron is shown above as 2×10^8. The ratio of these two numbers is 2×10^{-4}. So the mass of the rhotons taken from an electron is only two ten-thousandth of the electron mass.

To get an idea of the density of rhotons in the radiated wave front, I made some simplifying assumptions. Referring to the Figure 13 above, showing a dipole radiation pattern, assume that point A is at a distance of one wavelength, 3 meters. The circumference of this vertical circle is about 10 meters. Most of the radiation will be passing out through the semi-circle, 5 meters around.

The diameter of the horizontal circle will average between 3 and 6 meters. I will assume 4.5 meters. Therefore, the circumference of this circle is about 15 meters.

The total area through which the radiation will pass is the product of these two lengths, 75 square meters. By inspecting the derivation of the this area it is obvious that the area will increase by the square of the distance. Therefore the rhoton density will fall off as the inverse square of the distance.

The rhoton density, at any specific distance, will be the number of rhotons transmitted per cycle, divided by the area they pass through, at that distance. As an example; the density at the distance of 3 meters would be 2×10^{17} rhotons per sq. meter. A receiving antenna, placed at that point, would intercept the number of rhotons that would fall on the presented area of the antenna.

If the antenna were 1.5 meters long by .01 meter wide, its area, of .015 square meters, would intercept 3×10^{15} rhotons per cycle. Assume that each of the rhotons, that impinged on the antenna, were to free a valence electron. The total number of electrons that would flow, under the influence of the accompanying magnetic fields, would be less than a thousandth of a coulomb. Hence, less than a milli-ampere of current would be detected. Since, after each half cycle of a rhoton, the direction of its magnetic field reverses, the above current would reverse its direction of flow each half cycle.

HYDROGEN

Almost everything we see in the sky is radiation from hot hydrogen atoms in the outer regions of stars. A very small amount of the light comes from the other trace elements. To understand how this happens we must consider the hydrogen atom.

The electron spins around the nucleus (a proton) at a relatively large distance, compared to the size of the nucleus. The atoms in the gas interact with each other more frequently with increased temperature. This raises the energy of the electrons in some of the atoms. The electrons return to their normal energy levels by emitting photons of light. It was found that the photons emitted from hydrogen atoms had a specific pattern of discrete wavelengths, as measured by a spectrometer. The radiation from other elements had their own distinctive patterns.

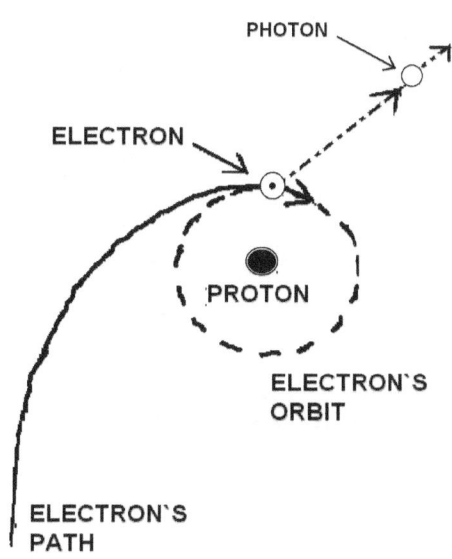

Figure 14

In Figure 14, the path of a free electron is shown as it approaches a positively charged hydrogen ion. As it nears the proton, the electron emits a photon of light and goes into orbit around the nucleus.

If this orbit happened to be the closest one to the proton that can occur, the photon emitted would have the largest momentum of any of the photons in the hydrogen spectrum. It would have a frequency that placed it in the ultra-violet region.

A theory about the hydrogen atom has been named The Bohr Model after the scientist who developed it. A representation of this is shown in Figure 15. This model has a number of nested orbits with systematically increasing radii. His theory states that electrons could only enter orbits with specific energy levels. The photons emitted would have momentum determined by the difference in energy levels between the starting and ending orbits.

Figure 15

In Figure 15 the free electron first entered one of the outer orbits and the energy difference was small. Therefore, the first photon was of low momentum and frequency. The second transition shown was into the innermost orbit and had a larger energy change.

The second photon had a larger momentum and higher frequency than the first. Interestingly, the sum of these two photon energies equaled the energy of the photon shown in Figure 14.

Some electron transitions from various orbits going directly to the innermost orbit produce a spectrum with numerically related frequencies. This spectrum is called the Lyman spectrum after the scientist who discovered it.

Similarly, some transitions that end at the next closest orbit to the nucleus produce a spectrum named after its discoverer, Balmer. The highest frequency photons in this series are at the blue end of the visible spectrum.

Why does an electron go into orbit around a proton instead of continuing on and colliding with it? When I learned of the quark theory of the make up of the proton, I began to see why the electron acted as it did. Figure 16, shows an electron going into orbit around a proton. Obviously, the spacing between the electron and the proton is not to scale. I show the proton according to my concept of how it may be configured.

Figure 16

The proton has two Up quarks, each with a +2/3 charge. They would tend to repel each other. It has been stated that something called gluons keep them from flying apart. Instead of gluons I assume that the negatively charged Down quark between the Up quarks can tend to hold them together.

The Down quark, with its −1/3 charge, will attract the two Up quarks, somewhat. The total charge of the proton is the sum of all the quark charges, a positive one.

One trivial point bothered me. Why the −1/3 and 2/3 charges of the quarks? Then I realized that the definition of a unit of charge is completely arbitrary. We could define the Down quark as −1 and the Up quark as +2. Then the proton would be +3 and the electron would be −3.

In addition to their charges, the quarks have spin. This produces a magnetic field around each quark. Assuming that each quark spins in the same direction, the Up quarks will have their positive (north) magnetic poles in the same orientation and the Down quark will have its positive pole in the opposite orientation. This will also cause the magnetism of the Down quark to attract both Up quarks and hold them close.

The entire proton also spins. Its electric and magnetic fields will interact with the approaching electron as they traverse it. The electron spins and has its own electric and magnetic fields. It will be attracted to the Up quarks positive electric fields but be repelled by the same orientation of its magnetic field with that of the Up quarks.

As the Down quark interacts with the electron, the like electric fields will repel while their oppositely oriented magnetic fields will attract.

At some orbital distance these forces should produce a critical interaction. I believe that it is during this interaction that the photon is produced.

The nature of an electron has not been given much consideration. I believe an electron may have a granular or fluidic composition. Under what could be considered a tidal interaction with a nucleus, an electron may throw off small bits of itself (photons), to achieve an orbit around the nucleus.

Consider the various factors that are involved in the electron and proton interaction.

If the velocity of the electron were small compared to the rate of rotation of the proton, successive periods of electric attraction and repulsion would impinge on the electron at the rate of the proton spin.

If the velocity of the electron were a significant factor, the lengths of the alternating periods would depend on whether the electron moved in the direction of the rotation of the proton, or in the opposite direction.

To simplify the analysis, assume that the electron does not travel far during a short time interval of interest. The four different electro-magnetic zones of the proton will traverse the electron each revolution of the proton, for a number of revolutions during this time interval.

The electron is also rotating. At some combination of all the factors of the interaction, the electron could be swept by two successive fields of the proton, for every rotation of the electron. The effect of this is shown in Figure 17.

Figure 17

When the Up quark is facing the electron, its positive charge attracts the closest surface of the electron, raising a bulge. By the time that the Down quark is facing the electron, the bulge has been rotated to the opposite side of the electron.

The repulsion of the negative charge of the Down quark, further increases the size of the bulge. This cycle will be repeated until the bulge becomes large enough to separate from the electron and fly off as a photon.

The interaction of the magnetic fields, of the quarks and the electron, could produce motions of the entire electron that are synchronous with the above interaction. This additional motion could enhance the building of the bulge and the separation of the photon.

Assume the photon is launched from the far side of the electron, while the Down quark is repelling the charge, and as the magnetic field of the quark is attracting the electron. The contributors to the photon velocity should be the speed and spin of the electron, and the repelling, negative electric field. I am assuming that the interaction with gravitons will cause the photon to achieve the speed of light.

The questions remain, how does this interaction cause the electron to go into orbit around the proton? Does more than one of these bits of charge get thrown off until a stable interaction is achieved? Does a photon consist of a series of charges, perhaps with magnetic fields that alternate? If this were the case, the spacing between the bits could determine the wavelength of the photon.

The proton could rotate around an axis that was at right angles to the one in Figure 17. This is shown in Figure 18 below.

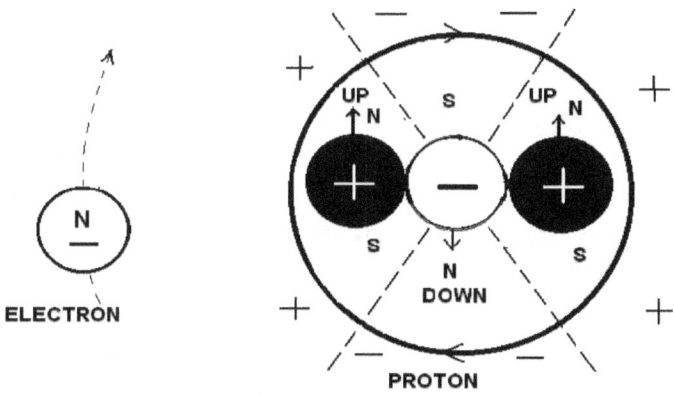

Figure 18

In this case, the magnetic fields of the quarks would not interact with the magnetic field of the electron, as shown. Only the electric fields of the quarks would alternately attract and repel the electron. This case might also tend to produce the bulge, and hence the photon. I think the contribution of the magnetic fields is needed.

I will assume that the spin rates of the electron and the proton do not change during this encounter. The main variables to consider are the velocity of the electron and the distance from the proton to the electron. I will also assume that the direction of the electron motion is the same as the direction of the sweep of the fields coming from the proton.

It seems that a stable orbit could come about when the electron is moving at the same velocity as the nucleus fields. The fields would no longer alternate across the electron and would not cause photons.

This would mean that the velocity of an electron in one of the larger, outer orbits would be greater than the velocity of one in a smaller, inner orbit. This is because each electron would have to make a single orbit in the time the proton rotates once. At any other electron velocity or distance, the effect of the alternating fields would depend on their rate of alternation, relative to the spin rate of the electron.

It can be seen in Figure 17, that two rotations of the electron are shown for every single rotation of the proton. Other odd multiples of this electron spin rate could also produce similar alternating interactions. Perhaps photons would occur with six rotations of the electron for each proton rotation, as well as for ten. Such a progression might occur as synchronous rotation is approached at different electron-proton spacing.

As a photon is emitted, the momentum imparted to the photon is subtracted from the momentum of the electron. This will tend to slow the electron. The electron will go to an orbit where it can revolve around the proton, in synchronism with the motion of the electric fields of the proton.

There are several conditions that could prevent an electron from going into orbit around a nucleus. The electron could approach too fast to be slowed sufficiently. It could be coming on the wrong side of the nucleus, and would have to circle against the sweep of the nuclear fields. The electron could come on the correct side but with the wrong direction of the electron spin polarity.

If an electron approached the nucleus from a direction along the nuclear spin axis, the nuclear fields would not affect it. It would respond to the net electric fields of the nucleus and the surrounding electrons. Such an electron would either be deflected or would continue on to impact the nucleus.

STARS

How does hydrogen, the smallest element, create stars? An electron spins around a single proton, the nucleus of the hydrogen atom. As previously shown, the proton is made up of even smaller components called quarks. It contains two "Up" quarks, each with a charge of +2/3. It also contains one "Down" quark with a charge of –1/3.

Figure 19 below, shows how a very high-energy electron can interact with a proton and produce a neutron. The electron can be thought of as having combined with one of the Up quarks, changing it to a Down quark. The resulting particle then contains two Down quarks and one Up quark. This is a neutron.

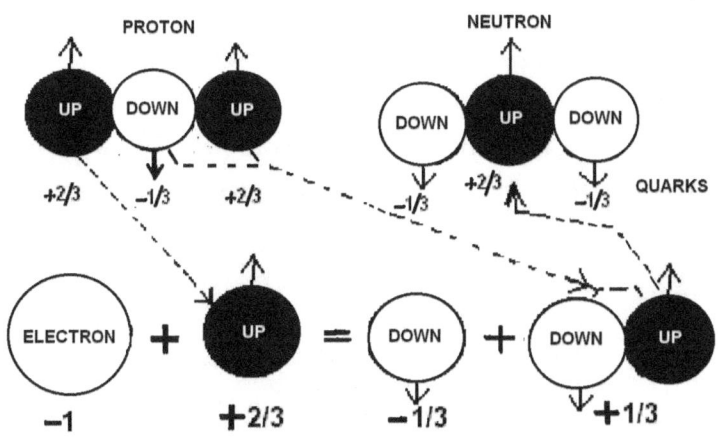

Figure 19

When a neutron combines with the nucleus of a hydrogen atom, it forms the isotope of hydrogen called deuterium, which is also a significant constituent of stars.

Figures 20 and 21 show two configurations of the deuterium nucleus. In Figure 20 only the electric fields would produce interactions with an electron. I do not think this would produce photons.

Figure 20

In Figure 21 below, the configuration would allow alternating interactions with an electron, from both electric and magnetic fields of the nucleus. The electron and the Up quarks must have the same magnetic field orientations.

Figure 21

An approaching electron would experience six alternations of the fields for each rotation of the deuterium nucleus. The electron would have to rotate three times, during a single rotation of the nucleus to produce the same kind of interaction as is shown in Figure 17 for the hydrogen atom.

I assume that odd multiples of this spin rate of the electron, relative to the deuterium atom, such as nine and fifteen, would also produce similar effects.

I should repeat, that it is not the actual spin rate of the electron that is different. It is the relationship between the nucleus spin rate and that of the electron that changes. Since the deuterium nucleus has twice the mass of the proton, it is reasonable that it would rotate slower than the proton. Then the required electron distance and velocity, in the two atoms, should be similar when they produce photons.

These two atoms, hydrogen and deuterium, were the first ones that formed, after the start of our universe. They constituted the greatest majority of the atoms in the universe. However, the amount of hydrogen was far greater than that of deuterium.

These atoms formed gasses that gathered into large, globular concentrations. This began the process of star formation. The weight of a globular volume of gas produced a high pressure at the center. It was the presence of the deuterium at the center that allowed an easier start of nuclear fusion. Two deuterium atoms combined to form an atom of helium, which has a slightly smaller mass than two deuterium atoms.

The net loss of mass turned into energetic, subatomic particles, which heated the core of the star. The increased heat exerted enough pressure to resist the inward pressure of the great mass of the star.

There is a rare reaction, involving a proton combining with a tritium atom that can occur. Tritium is a combination of a proton with two neutrons. However, the most prevalent energy producing reaction in the core of a star is when two deuterium atoms combined forming helium.

The structure of the helium atom would have to be compatible with handling two valence electrons. The electrons would need to have opposite spin axis polarities. What I show in Figure 22 is a possibility. It is essentially two deuterium atoms, with one flipped over and sandwiched with the other. I assume that this nucleus rotates slower than the deuterium nuclei did before.

Figure 22

The helium nucleus rotates in the same direction as the two electrons. One of the electrons will have interacted with one of the deuterium rings and produced one or more photons. The other electron has the opposite spin polarity and will have interacted with the other deuterium ring to produce photons. Both electrons will be in similar orbits, relatively close to the nucleus. The magnetic fields are essential for the correct interactions to occur.

Over an extended period of time, the supply of deuterium became depleted and the core temperature started to decrease. The weight of the star would compress the core further until the next stage of nuclear fusion started. Helium atoms started to form heavier atoms, with the generation of enough additional heat to stop further compression.

I have also tried to extend this concept to the lithium atom. Since it contains an additional proton and electron, I assumed an additional neutron. Therefore, I added another deuterium ring to the helium atom, with the results shown in Figure 23.

Figure 23

It seems that this configuration of lithium, with three protons and three neutrons, occurs a very small percentage of the time in nature.

The majority of lithium nuclei contain a fourth neutron. This extra neutron could be inside the nucleus, as shown in Figure 24. This should add to the stability of the nucleus.

Figure 24

The first two electrons of the lithium atom are in the innermost orbital region, as in the helium atom. The third electron must occupy an orbit farther out, reacting to the electric and magnetic fields that are the result of the alternating vertically oriented protons and neutrons. The electron (number 3) will be attracted by the net positive charge of a proton but repelled by their similar magnetic polarities. It will also be repelled by the effective net negative charge of a neutron but attracted by their oppositely oriented magnetic polarities.

I have included tables of the elements at the end of this chapter. These tables give more details about the various atoms. See pages 81 to 83.

The next elements, after lithium, shown on the tables are beryllium and boron. Each adds a single proton and neutron to the nucleus of the previous atom. Of course, they also have one more valence electron than the previous atom.

The following element, carbon, adds only a proton to the number in the boron nucleus. This evens up the number of protons and neutrons to six of each. This is called, carbon twelve. Two other isotopes of carbon exist, having one or two additional neutrons.

The next two elements, nitrogen and oxygen, each add a combination of a single proton and neutron to the previous nucleus.

The heavier elements all have increasing numbers of neutrons, in addition to the number of protons. I feel that there might be a physical picture of how atoms are built.

Based on the above information, I have reached some conclusions and have made the following assumptions:

1. As the atomic nuclei get larger, with more protons and even more neutrons, they must build up a fixed structure that exposes all the protons on the exterior surface of the structure. This provides the mechanism for controlling the orbiting electrons in their proper locations.

2. The protons, with neutrons interposed, are assembled into variously sized rings. This configuration will produce the alternating fields to interact with the orbiting electrons.

3. The surplus neutrons will accumulate within the rings and could assemble into a support and connecting structure for the rings.

4. The pattern of the electron numbers in the orbital shells will have been caused by the pattern of the protons in the associated rings of each element.

5. The fact that the electrons in each of the allowed orbits are in pairs, suggests that they have oppositely directed, magnetic polarities. I have assumed that the ring structure will also have pairs of rings with oppositely directed, positive quark polarities.

6. The larger the proton rings of the nucleus become, the larger their orbital electrons` spin rates should become to cause photon emission.

To illustrate these ideas, I have adopted a simplified way of drawing the protons and neutrons in the various sized rings, as rectangles with dark dots for Up quarks and lighter dots for Down quarks. It should be realized that the rings probably assume a circular shape, with alternating positive and negatively charged quarks. In Figure 25 the simplest 2-proton and 3-proton rings are shown in both formats.

Figure 25

Similarly, in Figure 26, I show the next rings, with increasing numbers of protons.

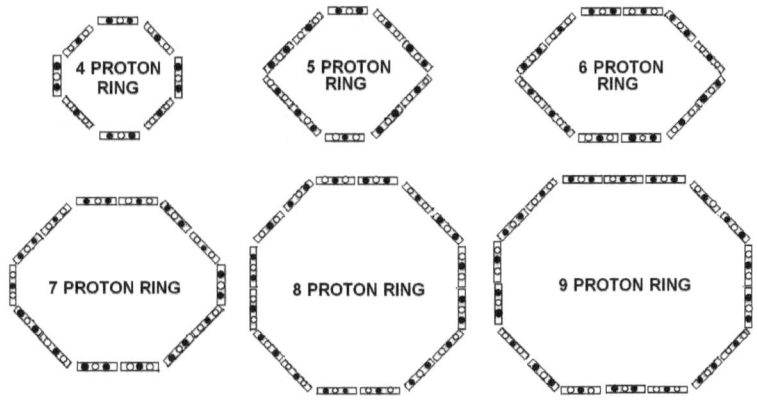

Figure 26

I am assuming that electrons in the closest orbits, controlled by the one proton rings, have a minimum of three times the spin rate of the nucleus when they emit photons. The larger rings should have electrons with spin rates that increase by a factor proportional to the number of protons in each ring. This should give these more distant electrons a stronger magnetic field to better interact with the nucleus and emit photons.

Illustrated in Figure 27, is an assembly of neutrons, having a linear and branching configuration. This could act as the central axis of the nucleus, with the branches connecting to the rings. The negatively charged Down quarks at the ends of the neutrons should be attracted to positively charged Up quarks.

Figure 27

How a ring would fasten to a neutron branch is shown in Figure 28. The center neutron would be part of the axial neutron string.

10 PROTON RING WITH 17 NEUTRONS

Figure 28

Pairs of rings with the same number of protons, but with opposed magnetic polarities, should hold together the same as the rings in the helium example. (See Figure 22.) But if there is too great a disparity in the proton number, from one ring to the next, some additional neutron bracing might occur, such as is shown below.

8 PROTON AND 16 PROTON RINGS

Figure 29

In Figure 29 a 16-proton ring is shown adjacent to an 8-proton ring. Four linear neutron strings connect the two rings together. It should be noted that the negative quarks at the ends of the neutron strings, connect to positive quarks at each end. I have included a side view of the rings, which I will discuss below.

Another simplifying technique used to help visualize the various atomic structures is to show the side view of a ring. It is shown as a vertical rectangle. The number of protons in each ring is given. This is shown in Figure 30.

BERYLLIUM 4 / 5

4 ⊟⊟⊟⊟
  1 1 1 1

BORON 5 / 6

5 ⊟⊟⊟⊟
  1 1 1 2

CARBON 6 / 6

6 ⊟⊟⊟⊟
  1 1 2 2

NITROGEN 7 / 7

7 ⊟⊟⊟⊟
  1 1 2 3

OXYGEN 8 / 8

8 ⊟⊟⊟⊟
  1 1 3 3

FLUORINE 9 / 10

9 ⊟⊟⊟⊟
  1 1 3 4

NEON 10 / 10

10 ⊟⊟⊟⊟
   1 1 4 4

SODIUM 11 / 12

11 ⊟⊟⊟⊟⊟
   1 1 4 4 1

MAGNESIUM 12 / 12

12 ⊟⊟⊟⊟⊟⊟
   1 1 4 4 1 1

Figure 30

This figure starts with the beryllium nucleus in the upper left corner. This follows the 3-proton nucleus of lithium, as shown in Figure 23. But in Figure 30, the axis has been shift to horizontal instead of vertical. Beryllium is indicated to have 4 protons and 5 neutrons, and is shown with four 1-proton rings.

The neon nucleus is shown with two 1-proton rings and two 4-proton rings. These rings control first and second electron shells that are completely filled. Thereafter, the first proton rings of all the elements are like this. Since this is such an important interface, I show in Figure 31 how the one and four proton rings can be tied together.

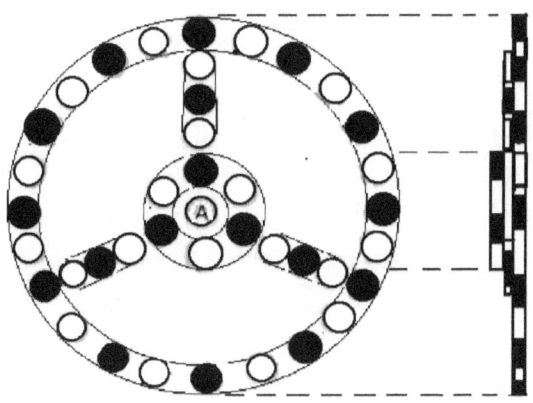

Figure 31

It can be seen that there are six quarks in the one-proton inner ring and twenty-four quarks are in the four-proton outer ring. Therefore, three neutrons are required to bridge between the corresponding pairs of positive quarks. An "A" indicates an axial quark in the center.

The next nucleus, in Figure 30 is sodium. It has a single proton ring added to the right end, controlling the valence electron in the highest orbit.

Magnesium has a second 1-proton ring added. The subsequent elements increase their number of protons in these rings until a total of eight is reached, with two, 4-proton rings. I assume that the following elements will increase the size of their proton rings according to the pattern of electron increases shown in Tables 1.

This sequence would repeat in the core of stars until the final fusions produced iron atoms. This was the beginning of the end for the star, since iron would not support the kind of fusion that produced energy. Many of these first generation stars would eventually explode and disperse the various elements into space where they would mix with the hydrogen clouds. From these clouds the second-generation stars were born.

In these stars the nuclear process in the core produced   neutrons that bombarded the heavier elements, such as iron. This produced even heavier elements over time.

For completeness, I am showing some of the heaviest nuclei in Figure 32. These are listed in Table 3.

Radon is the heaviest of the inert gasses. It shows the pattern of completed shells, from 2 to 8 to 18 to 32, without a final valence shell. Uranium is the heaviest, naturally occurring element. Plutonium is the familiar, man-made element.

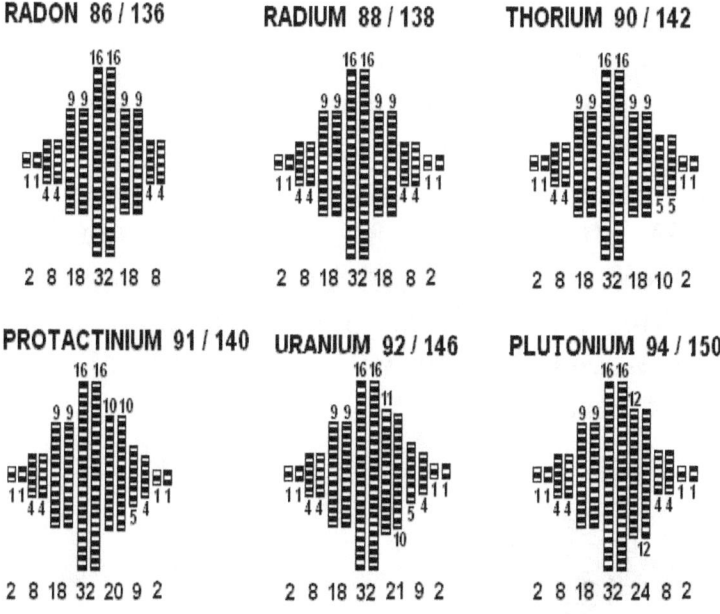

Figure 32

In all of these nuclei one can see how many additional neutrons are included, by noting the second number given. For example, in uranium, with 92 of the neutrons paired with the protons, the remaining 54 neutrons are inside the nucleus.

Note that in all of the above elements the maximum size rings occurs near the middle of the ring structure. They have 32-protons in a pair of 16-proton rings. The following rings get progressively smaller and control electrons in the higher orbits.

The electrons for the 16-proton ring should have a spin rate of about 48 times the nucleus spin rate when their photons are emitted. The electrons in the higher orbital shells should have spin rates at least this large.

A one-proton ring to control a valence electron in the highest orbit should have an electron with a spin rate much larger than the spin rates of the electrons in the lowest orbits. The greater spin of this electron should give it a larger magnetic field strength to interact with the nucleus.

For one last point of interest, I am including Figure 33. It shows the heaviest man-made nucleus that I have found. It is called unumbium and was made by smashing zinc and lead nuclei together. This is a very short-lived element with a half- life of about a quarter of a second.

2 8 18 2
ZINC 30/35

2 8 18 32 18 4
LEAD 82/125

2 8 18 32 32 18 2
UNUMBIUM 112/165

ZINC PLUS LEAD
MADE UNUMBIUM

Figure 33

It is interesting to note that the first four ring pairs of the lead nucleus became the front of the unumbium nucleus without change. Also, the last two ring pairs of the zinc nucleus became the back end of the unumbium. All the other parts of the lead and zinc nuclei went into the additional pair of 16-proton rings of the unumbium.

This picture of atoms presents some questions to be considered, such as, "What makes the nucleus spin?" Remember that heavier atoms were built up, in the stars, by adding helium, deuterons and neutrons to lighter atoms. As the heavier, ionized atoms attracted electrons, some of the angular momentum, that the electrons gave up to go into orbit, was added to the angular momentum of the nuclei. This increased the spin of the nuclei.

Another question is, "How do the new electrons know where to go into orbit?" Modifying an atom, which already had its full complements of electrons, made a new ionized atom. An electron approaching the nucleus of this ionized atom, would find that only the ring with the added proton was lacking an electron. The new electron would avoid the other electrons and respond only to this ring, until a stable, synchronized, orbit was achieved.

All the electrons in an atom can be visualized as being in a flat halo, circling the nucleus in unison. Each electron would tend to stay in its own place, in a relatively stationary orbit over its own proton ring.

Any tendency of an electron to deviate from this stability would be corrected by its interaction with the multitude of electro-magnetic fields. The above concept gives a consistent, logical approach to understanding how atoms may be structured.

TABLE 1

	Elements	Atomic Nuclei		Isotope Neutrons	Electrons in Orbital Shells					
		Protons	Neutrons		No.1	No.2	No.3	No.4	No.5	no.6
H	hydrogen	1	0	1 and 2	1					
He	helium	2	2	1	2					
Li	lithium	3	4	3	2	1				
Be	berylium	4	5	6	2	2				
B	boron	5	6	5	2	3				
C	carbon	6	6	7 and 8	2	4				
N	nitrogen	7	7	8	2	5				
O	oxygen	8	8	9 and 10	2	6				
F	fluorine	9	10	9	2	7				
Ne	neon	10	10	11 and 12	2	8				
Na	sodium	11	12	11 and 13	2	8	1			
Mg	magnesium	12	12	13 and 14	2	8	2			
Al	aluminum	13	14	13	2	8	3			
Si	silicon	14	14	15 and 16	2	8	4			
P	phosphorus	15	16	15 and 17	2	8	5			
S	sulfur	16	16	17,18 and 20	2	8	6			
Cl	chlorine	17	18	20	2	8	7			
Ar	argon	18	22	18 and 20	2	8	8			
K	potassium	19	20	22	2	8	8	1		
Ca	calcium	20	20	22 to 28	2	8	8	2		
Sc	scandium	21	24	23 to 28	2	8	9	2		
Ti	titanium	22	26	24 to 28	2	8	10	2		
V	vanadium	23	28	25 to 29	2	8	11	2		
Cr	chromium	24	28	26, 29 and 30	2	8	12	2		
Mn	manganese	25	30	28	2	8	13	2		
Fe	iron	26	30	28, 31 and 32	2	8	14	2		
Co	cobalt	27	32	29 - 34	2	8	15	2		
Ni	nickel	28	31	28 - 37	2	8	16	2		
Cu	copper	29	35	32 - 37	2	8	17	2		
Zn	zinc	30	35	32-42	2	8	18	2		
Ga	gallium	31	39	35-41	2	8	18	3		
Ge	germanium	32	41	34-43	2	8	18	4		
As	arsenic	33	42	38 - 45	2	8	18	5		
Se	selenium	34	45	39 - 47	2	8	18	6		
Br	bromium	35	45	41-50	2	8	18	7		
Kr	krypton	36	48	41-54	2	8	18	8		
Rb	rubidium	37	48	44 - 53	2	8	18	8	1	
Sr	strontium	38	50	47 - 56	2	8	18	8	2	

TABLE 2

	Element	Atomic Nuclei			Electrons in Orbital Shells						
		Protons	Neutrons	Isotope Neutrons	No. 1	No. 2	No. 3	No. 4	No. 5	No. 6	No. 7
Y	yttrium	39	50	47-56	2	8	18	9	2		
Zr	zirconium	40	51	47-58	2	8	18	10	2		
Nb	niobium	41	52	49-56	2	8	18	12	1		
Mo	molybdenum	42	54	49 - 59	2	8	18	13	1		
Tc	technetium	43	55	52 - 57	2	8	18	14	1		
Ru	ruthenium	44	57	52 - 62	2	8	18	15	1		
Rh	rhodium	45	58	56 - 61	2	8	18	16	1		
Pd	palladium	46	60	56 - 64	2	8	18	18			
Ag	silver	47	61	57 - 64	2	8	18	18	1		
Cd	cadmium	48	64	59 - 70	2	8	18	18	2		
In	indium	49	66	63 - 68	2	8	18	18	3		
Sn	tin	50	69	62 - 76	2	8	18	18	4		
Sb	antimony	51	71	67 - 78	2	8	18	18	5		
Te	tellerium	52	76	67 - 82	2	8	18	18	6		
I	iodine	53	74	69 - 83	2	8	18	18	7		
Xe	xenon	54	77	68 - 84	2	8	18	18	8		
Cs	cesium	55	78	74 - 84	2	8	18	18	8	1	
Ba	barium	56	81		2	8	18	18	8	2	
La	lanthium	57	62		2	8	18	18	9	2	
Ce	cerium	58	82		2	8	18	20	8	2	
Pr	praseodymium	59	82		2	8	18	21	8	2	
Nd	neodymium	60	84		2	8	18	22	8	2	
Pm	promethium	61	84		2	8	18	23	8	2	
Sm	samarium	62	88		2	8	18	24	8	2	
Eu	europium	63	89		2	8	18	25	8	2	
Gd	gadolium	64	93	85 - 97	2	8	18	25	9	2	
Tb	terbium	65	94		2	8	18	27	8	2	
Dy	dysprosium	66	97		2	8	18	28	8	2	
Ho	holmium	67	98		2	8	18	29	8	2	
Er	erbium	68	99		2	8	18	30	8	2	
Tm	thulium	69	100		2	8	18	31	8	2	
Yb	ytterbium	70	103	98 - 106	2	8	18	32	8	2	
Lu	lutetium	71	104		2	8	18	32	9	2	
Hf	hafnium	72	107		2	8	18	32	10	2	
Ta	tantalium	73	108		2	8	18	32	11	2	
W	tungsten	74	110		2	8	18	32	12	2	
Re	rhenium	75	111		2	8	18	32	13	2	
Os	osmium	76	114		2	8	18	32	14	2	
Ir	itidium	77	115		2	8	18	32	15	2	
Pt	platinum	78	117		2	8	18	32	17	1	
Au	gold	79	118	115 - 120	2	8	18	32	18	1	
Hg	mercury	80	121		2	8	18	32	18	2	

TABLE 3.

	Elements	Atomic Nuclei		Isotope Neutrons	Electrons in Orbital Shells						
		Protons	Neutrons		No. 1	No. 2	No. 3	No. 4	No. 5	No. 6	No. 7
Tl	thallium	81	123		2	8	18	32	18	3	
Pb	lead	82	125		2	8	18	32	18	4	
Bi	bismuth	83	125		2	8	18	32	18	5	
Po	polonium	84	125		2	8	18	32	18	6	
At	astatine	85	125		2	8	18	32	18	7	
Rn	radon	86	136	130 -135	2	8	18	32	18	8	
Fr	francium	87	136		2	8	18	32	18	8	1
Ra	radium	88	138	134 - 139	2	8	18	32	18	8	2
Ac	actinium	89	138		2	8	18	32	18	9	2
Th	thorium	90	142		2	8	18	32	18	10	2
Pa	protactinium	91	140		2	8	18	32	20	9	2
U	uranium	92	146	138 - 148	2	8	18	32	21	9	2
Np	neptunium	93	144		2	8	18	32	23	8	2
Pu	plutonium	94	150		2	8	18	32	24	8	2
Am	americium	95	148		2	8	18	32	25	8	2
Cm	curium	96	151		2	8	18	32	25	9	2
Bk	berkelium	97	150		2	8	18	32	26	9	2
Cf	californium	98	153		2	8	18	32	28	8	2
				Half Life							
Es	einsteinium	99	153	1 yr.	2	8	18	32	29	8	2
Fm	fermium	100	157	3 min.	2	8	18	32	30	8	2
Md	mendelevium	101	157	50 sec.	2	8	18	32	31	8	2
No	nobelium	102	157	2 min.	2	8	18	32	32	8	2
Lr	lawrencium	103	159	13 sec.	2	8	18	32	32	9	2
Rf	rutherfordium	104	157	5 sec.	2	8	18	32	32	10	2
Db	dubnium	105	157	4 sec.	2	8	18	32	32	11	2
Sg	seaborgium	106	157	.3 sec.	2	8	18	32	32	12	2
Bh	bohrium	107	157	.1 sec.	2	8	18	32	32	13	2
Hs	hassium	108	157	2 ms.	2	8	18	32	32	14	2
Mt	meiterium	109	157	3 ms.	2	8	18	32	32	15	2
Uun	ununnilium	110	159	10 ms	2	8	18	32	32	17	1
Uuu	unununium	111	161	____	2	8	18	32	32	18	1
Uub	unumbium	112	165	.3sec.	2	8	18	32	32	18	2

BLACK HOLES

Our Sun is a second-generation star. It is one of the smaller sized stars. It is used to compare the size of other stars. We call our Sun "one solar mass". The smallest stars are about one half a solar mass and the largest are more than one hundred solar masses in size.

The amount of nuclear fusion needed to withstand the weight of a star varies with the size of the star. Our Sun has enough fuel to burn for billions of years. The largest stars may have to use up all their fuel in less than a million years. When a star has used up its fuel, a number of different things can happen. Some stars become large cool red stars. Some become small hot dwarf stars. Stars that are in the range of ten solar masses can explode and then collapse, becoming a neutron star.

The most spectacular event occurs when a star of about one hundred solar masses explodes as a super nova. It may then collapse into a black hole. Mathematicians have done much of the theoretical work on black holes. Their equations for the characteristics of matter at the end of the collapse of a star into a black hole, predict that the size of the body goes to zero and the density becomes infinite. However, these bodies are said to have a finite mass. This is called a singularity. I think this concept is hard to believe. I prefer one that is simpler and more logical.

The nature of neutron stars may be the best place to start in understanding my idea about black holes. Neutron stars are thought to have a surface shell of iron atoms, which are the remnants of the depleted nuclear core.

Inside of the surface shell are the neutrons, essentially in contact with one another. In the collapse of the core the electrons of the heavier atoms are forced into the nuclei. The electrons and protons combine, forming neutrons.

It is theorized that at the center of the neutron star some exotic particles could be forming. I wondered what these exotic particles could be. This led to the idea of neutral particles, made up the next heavier quarks, becoming the predominant component at the center of the individual black hole from a collapsed star. This particle should be a hundred times more massive than a neutron.

The black hole should still have a layered structure with an outside layer of iron. Within that there should be a layer of neutrons and inside of that there should be the particles made up of the charm and strange quarks. I really need a shorter name for those particles, such as "charstrons", perhaps.

Neutron stars are said to result from the collapse of stars of about eight solar masses. After they collapsed, they had masses of a little over one solar mass. I am assuming that this was essentially the mass of their nuclear core. The remainder of their original mass must have been mostly hydrogen gas blown off in the super nova. I understand that these bodies ended up with a diameter of about ten miles.

A scientist named Schwartzchild derived an expression that gave the relationship between a body's mass and the distance from which light could no longer escape.

For a one solar mass neutron star with a diameter of about ten miles, the Schwartzchild limit is a four-mile diameter sphere. This would be within the body of the neutron star. Therefore, the neutron star should still be visible.

In the following figure the relative size of different mass stars is shown. Their masses increase with their volumes, and their volumes increase as a function of their diameters cubed.

STAR SOLAR MASSES

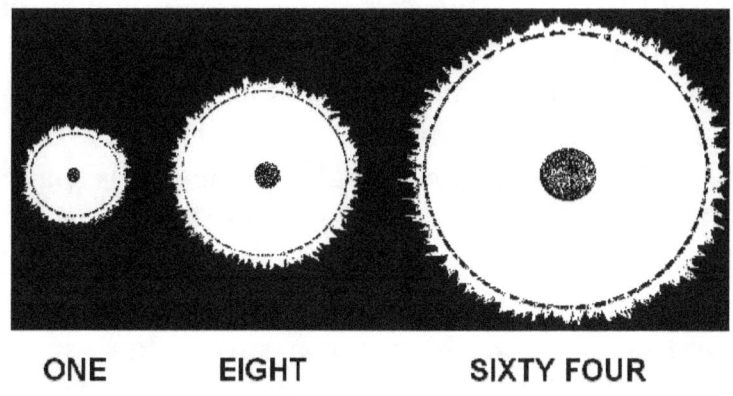

ONE   EIGHT   SIXTY FOUR

Figure 34

Black holes are supposed to occur some of the time when larger stars, in the range of fifty to one hundred solar masses, collapse in a super nova. I am assuming that a sixty-four solar mass star would have a core of about eight solar masses. If it had collapsed into a neutron star, which it might in a transient phase, it would have had a diameter twice that of the above neutron star. That is about a twenty-mile diameter. It should still have about the same mass of eight solar masses.

For this mass the Schwartzchild limit would be a sphere with about a thirty-two mile diameter. Therefore if this neutron star remained a neutron star it would be close to being a black hole.

However, assume that many of the neutrons in the center of this body were converted to charstrons. If the ratio, of neutrons being converted to charstrons, were about a hundred to one, the twenty-mile diameter of the eight solar mass body, would be reduced by a factor of about four. That is assuming the individual charstrons were about the same size as neutrons. It would stop collapsing at about a diameter of five miles. This would definitely be a black hole.

In the above examples, I have assumed that the Schwartzchild equation is valid. Since it is partially based on an Einstein theory it may not be.

If this first kind of black hole were in a situation where it was absorbing more and more matter from adjacent stars, it might reach a state where it would collapse again. A large number of "charm" and "strange" quarks would be converted into a smaller number of the more massive "top" and "bottom" quarks. These quarks would become neutral particles about one hundred times as massive as the charstron particles. Perhaps, we could call them "tobotrons". They might reduce the diameter of a charstron black body by a factor of four in becoming an equivalent tobotron black body.

I am assuming that it would be configured with a layer of iron on the outside. Within that there should be a layers of neutrons and charstrons.. The tobotrons would form within the layer of charstrons.

Perhaps a different name, other than black holes, would be appropriate. Astronomical bodies are commonly named using the suffix, "oid". That is like asteroid, meteoroid and planetoid. I would like to suggest that neutron stars be called "neutroids", charstron black holes be called "charstroids" and tobotron black holes be called "tobotroids".

The particles that would result from the conversion of quarks of one type to those of another are shown in the following figure. There is the possibility that the charstron quark charges are 100 times those of the individual neutron quarks. Also, this would imply that the charges of the tobotron quarks could be 100 times that of the charstron quarks and 10,000 times that of the neutron quarks.

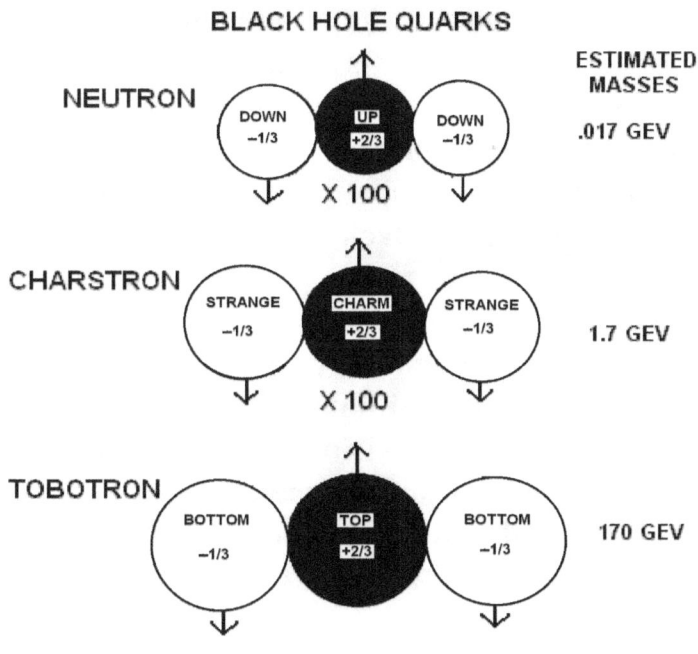

Figure 35

89

It is interesting to note that the above tobotron has about the same energy equivalent mass as that attributed to the Higgs particle, 170 GEV.

There is evidence that there are massive black holes at the center of most galaxies. They appear to have masses in the order of millions of times that of our Sun. This suggests to me that these could be black holes of the second kind (tobotroids). If a hypothetical tobotroid of ten solar masses would have a diameter of about one mile, a ten-million solar mass tobotroid, in the center of the galaxy, would have a diameter of about one hundred miles.

Assume that there are a hundred billion solar masses of stars in the galaxy. If all were ingested into this central black hole it would have a diameter increased by a factor of the cube root of 10 thousand, which is about 22. The final diameter of the galaxy weight black body would be about 2200 miles. (We could call this a galaxoid.)

If a hundred billion such galaxoids combined into a single massive black hole, its diameter would have increased by a factor of the cube root of a hundred billion (about 4500). Therefore the final diameter of a universe weight black body (univoid?) would be about ten million miles. This supposedly accounts for the present visible universe. Since most of the gravitons will remain dark, assume that they represent 7/8 of the total volume. This suggests a final diameter of twenty million miles.

It is conceivable that even more massive quarks could form at the center of such a body. These could create another type of neutral particles out of tobotrons. This could reduce the diameter of the universal black body even further.

Consider how black holes will interact with other matter. Gravitons will impact on a black hole from all directions. The gravitons will enter through the spaces between the surface atoms and make innumerable elastic collisions within, transmitting their momentum into the black hole.

Any material body in the vicinity would be accelerated toward the black hole, since no gravitons would be coming out of the black hole to impede them. The approaching bodies would achieve exceedingly high velocities and be assimilated into the black hole upon impact.

When black holes are in the same vicinity, they will tend to shield each other from gravitons coming from the directions beyond them. These gravitons will impart momentum to the black holes they fall upon and cause the black holes to move towards each other, until they merge.

It is evident that black holes would accumulate gravitons. These gravitons will have come from their own universe, from adjacent universes, and even from previous universes that no longer exist. The gravitons will be recycled, liberated when the black holes containing them turn into a Big Bang.

BIG BANG

We can look at the big bang in a new way. Instead of assuming a collapse of the universe size black hole, assume that something causes it to open up. Perhaps the impact of a large incoming mass (another black hole) could cause the black holes to crack open.

Another possibility is that if the flux of incoming gravitons dropped below a certain level, the internal pressure could crack the surface layers. This would release the charstrons and the tobotrons from containment. They would erupt in all directions.

The individual tobotrons could become a large number of charstrons. The charstrons could become many neutrons. These would race off in every direction at a velocity, greater than the present speed of light. This expanding sphere of high velocity neutrons would have enumerable collisions occurring.

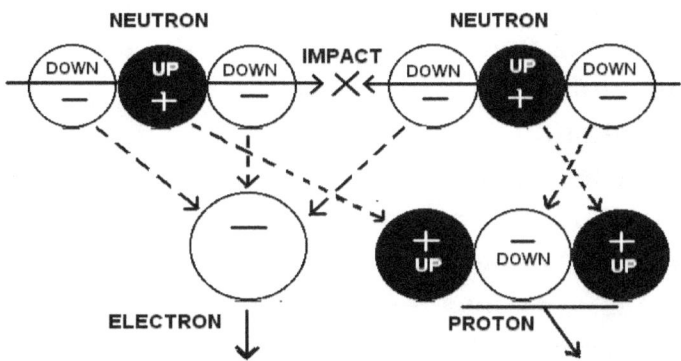

Figure 36

Figure 36 shows how a collision between two neutrons could produce a proton and an electron. Later, neutron collisions with protons could produce deuterons.

Eventually, electrons will combine with protons and become hydrogen. Also, electrons and deuterons will form deuterium. Star formation could then begin.

A soup of super-luminal neutrons could be filling the universe to this day. They can be entering all objects from all directions. They could make occasional impacts on the atoms of matter imparting momentum. These particles could be what I have called gravitons.

When I described how black holes accumulated gravitons, it could have been neutrons being accumulated. This would have increased the conversion of neutrons to charstrons and eventually increasing the number of tobotrons.

At the instant of the big bang, if all of the particles in the twenty million mile diameter black body turned instantly into neutrons, essentially in contact, the volume would have expanded by a factor of ten thousand. The diameter would be almost five hundred million miles.

After a relatively short period of this high concentration of neutrons, the density would rapidly decrease. At the present time the contribution of the big bang gravitons may only be a small portion of the total number of gravitons. Probably, the gravitons from the surrounding universes make up the rest.

GRAVITONS

At the instant of the big bang, the first force that is supposed to have emerged is gravity. It has been assumed that this was a force of attraction. But by my assumption, it was a repellant force and would initiate rapid expansion.

There has been an attempt to replace the "Inflationary Theory" in the Big Bang. One of the ideas is that at the very beginning of the universe, the velocity of light was far higher than it is now. With my concept of the speed of light being controlled by the speed of gravitons, this would imply that the speed of gravitons, at that time, was much higher than it is presently. It may be that over a long period of time the multiple collisions made by the gravitons would reduce their momentum slightly. This would cause a reduction in their velocities.

With further cooling, (reduced velocity impacts), the other forces would emerge, and then the primary particles would form. These particles would interact with the gravitons. Most of the gravitons would be scattered by the particles and remain among them.

As the universe evolved, the greatest portion of the visible mass of the universe would be in the central part of the distribution, at some average radius. Within this region, bodies would be impacted by gravitons coming from every direction. Only in the innermost and the outermost parts of the population would there be an imbalance of the graviton impacts.

Bodies in the innermost region would tend to be retarded and their rate of expansion would be slowed by the predominant direction of the graviton flux being from the central region.

The bodies that are in the farthest out region will also be affected by the graviton flux in an unbalanced way. They will be accelerated by the gravitons coming from the central region behind them.

What gravitons are is a good question. They would have to perfuse our space but not have been detected, except for the effect of gravitation they produce. Originally, I thought that the theoretical particle, the Higgs Boson was the graviton. If a large number of tobotrons had survived the big bang, they could be an equivalent of the Higgs particle and serve as gravitons.

The vastly more numerous neutrons that were released in the big bang are much better candidates to be gravitons. I have assumed that gravitons are neutrons traveling with velocities greater than the speed of photons.

What could be the interaction of a graviton with a photon? What could the interaction of the graviton with an atom of matter be like?

The graviton (neutron) has two negative, Down quarks linked by a positive Up quark. The Down quarks are in a quasi-stable configuration and will tend to oscillate and then cause the graviton to spin.

The following figure shows how the electromagnetic fields surrounding the spinning graviton might interact with photons while passing them.

Figure 37

The rotating electric fields of the graviton would be predominately from the negative quarks. The graviton is moving to the right with the velocity Vc+. The upper photon is moving to the left with a velocity Vc. The vector Vg shows the effect of the electric field of the graviton upon the negatively charged photon. The magnitude and direction of the resultant velocity of the photon is shown as Vr. The photon has been slowed down and deflected away from the graviton.

At a different time than that of the above encounter, the lower photon is traveling with the velocity Vc in the same direction as the graviton. As the graviton approaches the photon the effect of the electric field of the graviton is to accelerate the photon and deflect it away.

Since the mass of the graviton is so much greater than that of the photon, the velocity and direction of the graviton is not affected.

Because the graviton passes the lower photon with less of a relative velocity, it has more time to accelerate this photon. There is less time to decelerate the upper photon in their shorter duration encounter.

When a graviton approaches an atom of some heavier element the interaction is different. Occasionally a graviton may impact a nucleus. A graviton is more likely to be deflected by the negative charges of the orbiting electrons of the atom. The momentum of the graviton is so high that only the small amount of its momentum that is transferred to the atom will move the more massive atom a small amount, in this elastic collision. The graviton may not have to come very close to an atom to transfer some momentum to it.

It is now believed that about three quarters of the mass of the universe is dark. Dust clouds, black holes or neutrinos cannot explain this mass. It is also thought that something called dark energy exists which acts like a negative force of gravity. These ideas have been theorized because of observations of unexpected occurrences in distant super novas and galaxies.

My theory, that swarms of "faster than light" neutrons fill the universe and provide the effects of negative gravity, explains this. The graviton theory is consistent with the idea of dark matter. Neutrons do not absorb or emit light. They probably were the main product of the big bang.

However, the most likely source of most of the gravitons in our universe could be from outside our universe. Most of the mass in the universe could be gravitons.

A galaxy will tend to act as a sink for gravitons in that its black holes will absorb the gravitons. More gravitons will move towards a galaxy than will be moving away from it. This will cause a gradient in graviton flux density, which will cause the observed motions.

Starlight from more distant galaxies will be deflected by the graviton flux gradient around a galaxy. The path of light will curve from the denser graviton flux toward the less dense flux. This has been called a gravitational lens.

If one wanted to travel from Earth to some other star, it should be easier to reach one that is closer to the center of our galaxy, than one in the opposite direction. Our galaxy's graviton flux gradient should favor this.

MOTION

What is the effect of gravitons upon a body in motion? Assume that a small body is alone in space, and that it is in an unknown graviton flux. The body could be in a non-uniform graviton flux. There could be more gravitons coming from one direction than coming from another. Also, The velocities of the gravitons coming from one direction could be different than those coming from another.

The body would be accelerated by the gravitons, toward the source of the lower graviton flux or toward the source of the lower graviton velocities.

If the body is moving with a constant velocity, relative to some distant frame of reference, it is in a non-uniform graviton flux that compensates for the velocity of the body.

As long as the gravitons coming from any direction impart the same momentum to the body, as those coming from the opposite direction, there will be no change in the body's momentum.

This suggests that one of Newton's laws of motion is only approximately correct. As long as the body has a velocity that is small compared to the velocity of gravitons, it will appear to remain in motion without change.

With a source of thrust, the body can accelerate and achieve a higher velocity. This will give the body an increased velocity relative to the gravitons that impact the front of the body. They will transfer more momentum to the body than the gravitons impacting the rear. This imbalance will provide an opposing force that the thrust must overcome.

If the thrust is discontinued, the body will decelerate until it reaches a velocity, where it is balanced by the graviton flux differential.

The momentum of a neutron, with the velocity C, impacting the front of a body that is traveling at a velocity V, is the mass of a neutron (Mn) times (C+V). The momentum of a neutron impacting the rear of the body depends on (C-V). The net momentum, causing deceleration, is the difference between these, that is:

$$[(C+V)- (C-V)] (Mn) = 2V (Mn).$$

If sufficient thrust were applied for the body to reach a velocity that approached the speed of light, the momentum of the gravitons impacting the front would become twice as great and the momentum of those impacting from the rear would approach zero. It would take a very large thrust to maintain this velocity.

The required force for the body to continue accelerating, beyond the speed of light, will depend more on the increase in the momentum of the body than on the impacts of gravitons against it. Therefore with sufficient thrust a vehicle should be able to attain faster than light travel.

This is in direct contradiction to Einstein's theory of Special Relativity which states that an infinite amount of energy is required to accelerate a body to the speed of light. Because of this, it is important to either verify or refute my new theory of gravity and motion, for the benefit of man's future progress in space.

FASTER-THAN-LIGHT

This is a highly speculative idea that follows up on my suggestion that faster than light (FTL) travel may be possible. As I indicated in the chapter on motion, in order to maintain a given velocity, a vehicle will have to provide a constant thrust of the proper magnitude. To do this for inter-stellar travel, the vehicle must have a continuous supply of fuel. My idea is to tap the gravitons of space to provide this fuel.

I hope that a method can be developed that will cause, under controlled conditions, the kind of collisions between FTL neutrons that is described in the chapter on the Big Bang. (See Figure 36.) The protons and electrons produced will be the inputs to proton accelerators. The following diagram shows a possible configuration of a FTL propulsion system.

Faster Than Light Propulsion System

Figure 38

The main engineering challenge will be to concentrate the neutrons going into the collisions so as to provide adequate output protons.

As indicated, the vehicle is moving upward with a vertical velocity, V. The graviton concentrators are shown in a position appropriate for a V / C ratio of 1.

The horizontal components of the graviton flux will enter the concentrators from the right and the left. With the ship's velocity equal to the graviton velocity, the gravitons will follow the 45 degree path to the apex of the concentrators. I will call them horns hereafter.

With the ship starting out at a low velocity the horns will be in the position shown for V / C = .1 ratio. With increased velocity the horns will be moved forward as required. Eventually, a vehicle velocity of twice the speed of light may be reached with the horns in the position indicated as V / C = 2. The horns may be oriented separately but will usually be placed symmetrically. They will be capable of moving through arcs of almost 180 degrees.

A second pair of horns and another proton accelerator has been included in this configuration. This doubles the propulsion with only a little additional ship mass.

The ability of a horn to concentrate all the gravitons entering the aperture into a narrow beam at the collision point is critical. I can think of several possible approaches to accomplish this. The horns might have a passive design that uses their shape and material characteristics to deflect the neutrons toward the vertex. It is more likely that an active approach will be needed.

This will probably involve various electric and magnetic fields. These might be steady state fields or require higher energy pulses.

The region of the graviton collisions will also require an innovated design. The area will have to be shielded. The protons will have to be contained and directed down the accelerators. The electrons will have to be separated out and directed into electro static fields or into power regeneration.

The above describes how the vehicle accelerates to FTL speed. For the ship to decelerate, reducing or discontinuing the thrust will cause a reduction in velocity due to the head-on impacts of the graviton flux.

After the vehicle has reached a relatively low velocity, if a more rapid deceleration is needed, the ship may be rotated so that the proton accelerators face the forward direction. Their thrust will then cause a more rapid deceleration. To provide the fuel for this activity the horns can be moved to the appropriate position near the proton accelerator end of the ship.

The various problems of life support, recycling water and regenerating oxygen have been worked upon already. I assume that the primary long-term power source will be nuclear. For this to be a feasible interstellar vehicle, a great amount of back up capacity and redundancy will be required. Trips of many years, even at FTL speeds, will be required to reach the closest star systems.

In the following diagram I show how combining duplicate versions of the basic vehicle propulsion will provide a degree of extra safety.

CREW QUARTERS

Four Proton Accelerators
Two Power Modules

Six Proton Accelerators
Three Power Modules

MODULAR FASTER THAN LIGHT VEHICLES

Figure 39

As shown, the habitable areas have been placed forward, away from the internal sources of radiation. The quarters for the crew may be separated into several compartments, each with control of one or more of the propulsion units.

If the technique of flipping the vehicle over, to be able to use thrust for deceleration, is done at very high velocities, the floors would become ceilings and the ceilings would become floors. If this is a requirement, the compartments could be designed to rotate 180 degrees around a horizontal axis, as this takes place.

The effects on the crew, of the acceleration and constant velocity of the vehicle and the probable forces relative to the Earth's force of gravity, are a concern.

The acceleration should be maintained at a tolerable level. At some high velocity the difference in graviton flux between the head-on and following gravitons, and its effect on the momentum of the vehicle and its occupants, should be equivalent to the force of gravity experienced on Earth. If this occurs at a velocity close to the speed of light, then the crew might tolerate the additional force as the velocity is increased further. At the speed of light the vehicle and its occupants will experience a maximum retarding force, since only head-on graviton impacts will occur.

After the speed of light is reached, increasing the velocity will increase the momentum of the ship and crew but not the momentum of the head-on gravitons. The effective retarding force on the ship and crew may actually decrease with increased velocity.

The increased momentum of a body in the vertical direction may restrict most motion to the horizontal plane. This would also apply to photons and electrons in electronic control systems.

If approaching the speed of light produces intolerable effects, the maximum velocity will have to be limited to some lower value. Hopefully, this would be at some appreciable fraction of the speed of light.

CONCLUSIONS

I have presented my ideas of a worldview that may not be as mathematically elegant as those of Albert Einstein, but hopefully they will be more readily visualized. I see things differently then those who use the Relativity equations unquestionably as the starting point for their extrapolations. My ideas may not all prove to be correct, but they have the beauty of simplicity and being understandable.

Einstein was able to develop his Special Relativity in less than a year. It took him ten years to develop General Relativity, to tie gravity in with his first approach. He then spent the rest of his life trying, unsuccessfully, to truly unify gravity with the other forces of nature. I am the most uncomfortable about his use of time as an independent variable in equations that predict hypothetical effects of velocity on time and the mass of objects.

In summarizing the ideas presented here:
I believe that the three dimensional view of objects moving out into a universe of non-expanding space is useful to overcome the dead end of the expanding space model, which predicts that every other body in space will eventually disappear from our view, moving away at a speed faster than light.

I believe that our universe is a naturally occurring process that must have a means of reproducing itself. That is why the idea of multiple universes surrounding ours is a reasonable possibility.

The idea that photons have mass is not really too new. It is only the idea that they are limited to a velocity determined by gravitons that is new.

I consider the idea of gravitons, acting as a repellant force that imparts momentum to objects and produces the action of gravity, is the most significant of the ideas presented here.

The concept of black holes being knowable as objects that are produced by a natural extension of known processes is a reasonable one. This ties together a progression; from a large star mass black hole, to a galaxy mass black hole, to an universe mass black hole, and to the Big Bang.

I conclude that gravitons are neutrons traveling with velocities greater than the speed of light. These gravitons could be the mysterious dark mass that has been theorized as constituting the vast majority of the mass of the universe.

I suggest using electro-magnetic coupling, instead of "gluons" to provide a useful configuration of quarks, within protons and neutrons. This led to my idea of how the interaction of a proton with an electron produces photons in a hydrogen atom and also that the photons are very small bits of the electrons emitting them. I also, derived a concept of how electrons and nuclei interact and establish the electron orbits in atoms. I describe a ring structure for the nuclei of all the elements that can account for the structure of the electron orbits.

Finally, in a detailed explanation of how gravitons might interact with matter, I present a new theory of motion that modifies Newton's law of motion. My theory also appears to indicate that faster than light travel may be possible.

This book presents what may be the long sought, unified theory of gravity and electro-magnetism.

(This is from a self-portrait by the author)

Lenard Metzger was born and raised in Rochester, New York. He has lived there ever since, except for the years spent in service and college. He received a degree in physics from Ohio State University.

He was a long time employee of Eastman Kodak Company. After retiring he returned to his early interests in art, science and writing.

www.ingramcontent.com/pod-product-compliance
Lightning Source LLC
Chambersburg PA
CBHW022058170526
45157CB00004B/1389